NATIONAL GEOGRAPHIC

School Publishing

SERIOUS SURVIVORS

PATHFINDER EDITION

By Susan Halko

CONTENTS

SERIOUS SURVIVORS

By Susan Halko

A deer peacefully grazes in the woods. Suddenly, its white tail flips straight up. It's a warning sign to other deer. A hungry coyote is nearby. The deer bolts. It runs through the woods, jumping over logs and streams. The coyote runs after it. The deer darts left, and then right, trying to lose the coyote. After a while, the coyote gets tired and gives up. The deer escapes. It survives!

Most animals can run or hide to escape an enemy. But plants are rooted in place. How can they survive attacks from plant-eating animals and insects?

You might think that plants are helpless because they are stuck in one place. Think again. Plants have some amazing ways to survive.

Some plants discourage animals from attacking in the first place. They might have sharp thorns or spines that turn away enemies. Or they might trick insects and animals into not attacking them.

Other plants fight back, using built-in defenses. Some plants even call for help! Let's find out about these serious survivors.

Self-protection. This plant can't run from danger, but it still has ways to protect itself.

3

BEWARE OF THORNS

Some plants convince hungry plant eaters not to attack. They are equipped with **mechanical defenses**, or parts such as spines, thorns, or prickles. These razor-sharp features can pierce an animal's skin.

The leaves of a holly plant have spines around the edges that can make it hard for animals to eat.

The sharp thorns on the stem of some raspberry plants send a clear message: "Beware. You will get hurt if you try to attack."

PLAYING TRICKS

Some plants trick their enemies into not eating them. The passionflower's worst enemy is the caterpillar. Caterpillars love to eat it up! When a butterfly looks for a spot to lay its eggs, it often lands on a passionflower. When the butterfly eggs hatch into caterpillars, the caterpillars eat the leaves.

The passionflower tricks the butterfly as protection against these future attackers. It has small, yellow bumps that **mimic**, or copy, the look of butterfly eggs. A butterfly sees the bumps and looks for another place to land.

Holly Plant

Fake eggs on a passionflower leaf

Real butterfly eggs on a leaf

Living Rocks. Pebble plants hide easily among rocks. Can you spot the plants?

When a living thing looks like something else to fool enemies, it's called mimicry. Another example of mimicry is found in the desert of South Africa. The pebble plant mimics the ground around it. It looks just like—you guessed it—a pebble. Desert animals crawl right over the plant, never knowing that they just passed up a satisfying snack.

TINY BODYGUARDS

Sometimes plants team up with insects to fend off plant eaters. For example, fierce ants live inside the hollow thorns of acacia trees.

The ants, in turn, act as tiny bodyguards. They protect the tree by stinging intruders that would munch on the tree. The tree also makes food for the ants.

Plant Protectors. These ants protect the tree against anything that comes near it, including people!

FIGHTING BACK

Stop for a moment and think about the coyote chasing the deer in the forest. What would have happened if the coyote had attacked a skunk? P.U.! The skunk would have fought back with its smelly spray.

Some plants use **chemical defenses** to fight back, too. They just don't spray them like a skunk. Instead, they hide toxins in their leaves or stems. **Toxins** are chemicals that are harmful to animals. They can make it harder for an animal to digest the plant. They can even kill the attacker.

Persimmon trees produce a chemical that can make its fruit taste bitter. Foxglove is a plant that contains toxins that are deadly to some animals.

Foxglove

Pool Cover. The "lid" on the pitcher plant keeps rain from overflowing its pool.

BUG-EATING PLANTS

The pitcher plant has an interesting way of getting rid of intruding insects—it eats them!

You see, the pitcher plant grows in marshes and bogs. The soil in these swampy places does not have all the nutrients the plant needs. So, the pitcher plant gets nutrients from insects.

The pitcher plant has a long tube that fills with water and enzymes. If an insect wanders off of the rim of the plant, it slips and falls down into the pool of liquid. The insect drowns, and the plant takes in a good meal. Some pitcher plants are big enough to eat a frog!

CALLING FOR HELP

Some plants call for help when they are under attack. They don't howl or shriek. They use chemicals to send their messages.

When a caterpillar chews on the leaves of a cotton plant, chemicals in the insect's saliva mix with chemicals in the plant. The result is an odor that attracts the caterpillar's enemy, the wasp.

The wasp comes to the plant's rescue. It lays eggs on the caterpillar. In a few days, the eggs hatch and eat the caterpillar!

Scientists have found that plants send a different odor to attract different enemies, depending on what kind of insect attacks.

LIFE CYCLE

Thorny or slippery, trickster or warrior—all plants grow and change. They go through stages of the **life cycle**. The pictures below review the life cycle of a plant.

1 The seed germinates. A small root begins to grow downward, and a shoot grows upward.

2 The plant grows.

3 The plant flowers.

4 Flowers are pollinated. Then they form seeds.

5 The plant releases seeds.

6 The plant dies.

WORDWISE

chemical defense: a way that a plant protects itself by using chemicals

life cycle: a pattern of how a living thing changes as it grows

mechanical defense: features of a plant's structure that help it protect itself

mimic: to copy something

toxin: a poisonous substance

SNEAKY PLANTS

To keep the cycle of life going, plants must do more than survive enemy attacks. They need to reproduce! Plants can't go looking for a mate like animals can. So how do they get their pollen to other plants?

They get sneaky. Some plants trick insects' sense of sight, smell, and touch. They lure insects to collect their pollen and take it with them.

The mirror orchid is one of these sneaky plants. It uses mimicry to attract pollinators. Pollinators are animals that spread pollen from one plant to another.

The blue part on the flower looks just like the sky reflecting on a female wasp's wings. When a male wasp sees it, he thinks he has found a mate. So he lands on the flower. Then he flies away and carries that orchid's pollen to another orchid flower.

Mirror Image. This wasp's wings look similar to the orchid's petals.

Mirror Orchid

DELICIOUS SMELLS

Imagine walking past a bakery and catching a whiff of fresh bread. Pretty hard to resist, right? That's how the bucket orchid smells to male bees.

A male bee can't resist the smell of a bucket orchid. The bee lands on the orchid and slips on its smooth surface. He falls into the orchid's deep bucket, where he becomes stuck in a pool of a liquid at the bottom.

To escape, the bee has to squeeze through a tight tunnel. As he crawls out, a pollen sac gets stuck on the bee's back.

Then the bee is tricked again by another bucket orchid of the same species, and the same thing happens. The bee falls into the bucket and has to climb out.

But this time the bee is carrying the pollen sac from another orchid. As the bee climbs out, the pollen sac gets trapped in the female part of the flower. The bee pollinates the orchid, and he doesn't even know it!

The Great Escape. A bee climbs out of the bucket orchid with a pollen sac on his back.

Surprising Smell. This skunk cabbage might look pretty, but it has a bad smell.

Offensive Odor. Some orchids have a rotten odor that attracts flies.

DISGUSTING SMELLS

Skunk cabbage smells bad—like dead animals. It doesn't use the strong odor to defend itself, though. It uses its stench to attract pollinators.

Its strong odor smells like good food to hungry flies and beetles. They can't resist. They swarm over from one flower to another, spreading pollen as they go.

STAYING ALIVE

Plants might not be able to run from an enemy or go looking for a mate, but they have amazing ways of surviving. They are one of nature's greatest success stories.

SERIOUS SURVIVORS

Find out what you learned about how plants survive.

1 What is an example of a plant's mechanical defense?

2 What is an example of a plant's chemical defense?

3 How do ants help protect an acacia tree?

4 How does mimicry help a plant discourage animals from attacking?

5 How does mimicry help a plant attract pollinators?